THE GEOLOGIC TIME SCALE

A chart showing the sequence, names, and ages of Earth's rock layers (oldest at the bottom)

Eon of time Eonothem of rock	Era of time Erathem of rock	Period of time System of rock**	Epoch of time Series of rock	Millions of years ago (Ma)	Some notable fossils in named rock layers
Phanerozoic	Cenozoic: (new life) Age of Mammals	Quaternary (Q)	Holocene	.0117	First *Homo* fossils, 70–100% extant mollusks[+]
			Pleistocene	2.6	
		Tertiary: Neogene (N)	Pliocene	5.3	First humans (Hominidae), 15–70% extant mollusks[+]
			Miocene	23	
		Tertiary: Paleogene (P_G)	Oligocene	34	More mammals than reptiles, <15% extant mollusks[+]
			Eocene	56	
			Paleocene	66	
	Mesozoic: (middle life) Age of Reptiles	Cretaceous (K)		145	Last dinosaur fossils: including *Tyrannosaurus rex*
		Jurassic (J)		201	First bird fossil: *Archaeopteryx*
		Triassic (Ŧ)		252	First dinosaur, mammal, turtle, and crocodile fossils
	Paleozoic: (old life) Age of Trilobites	Permian (P)		299	Last (youngest) trilobite fossils
		Carboniferous (C)*: Pennsylvanian (ℙ)		323	First reptile fossils
		Carboniferous (C)*: Mississippian (M)		359	First fossil conifer trees
		Devonian (D)		419	First amphibian, insect, tree, and shark fossils
		Silurian (S)		443	First true land plant fossils
		Ordovician (O)		485	First fossils of coral and fish
		Cambrian (€)		541	First trilobite fossils First abundant visible fossils
Proterozoic	Precambrian: An informal name for all of this time and rock.			2500	Oldest fossils: mostly microscopic life, visible fossils rare
Archean		Oldest fossils of visible life (stromatolites)		3500	
				4000	
Hadean		Acasta Gneiss, northwestern Canada		4030	
		Nuvvuagittuq greenstone belt, Quebec, Canada		4280	
		Zircon mineral crystals in the Jack Hills Metaconglomerate, Western Australia		4400	
	Oldest meteorites			4550	

*European name

**Symbols in parentheses are abbreviations commonly used to designate the age of rock units on geologic maps.

[+]Extant mollusks are mollusks (clams, snails, squid, etc.) found as fossils and still living today.

FIGURE 1.3 The geologic time scale. Absolute ages in millions of years ago (Ma) follow the International Commission on Stratigraphy, 2013. See their website for more detailed versions and recent updates of the international geological time scale (**http://www.stratigraphy.org/index.php/ics-chart-timescale**).

Example of Geologic Field and Laboratory Investigation

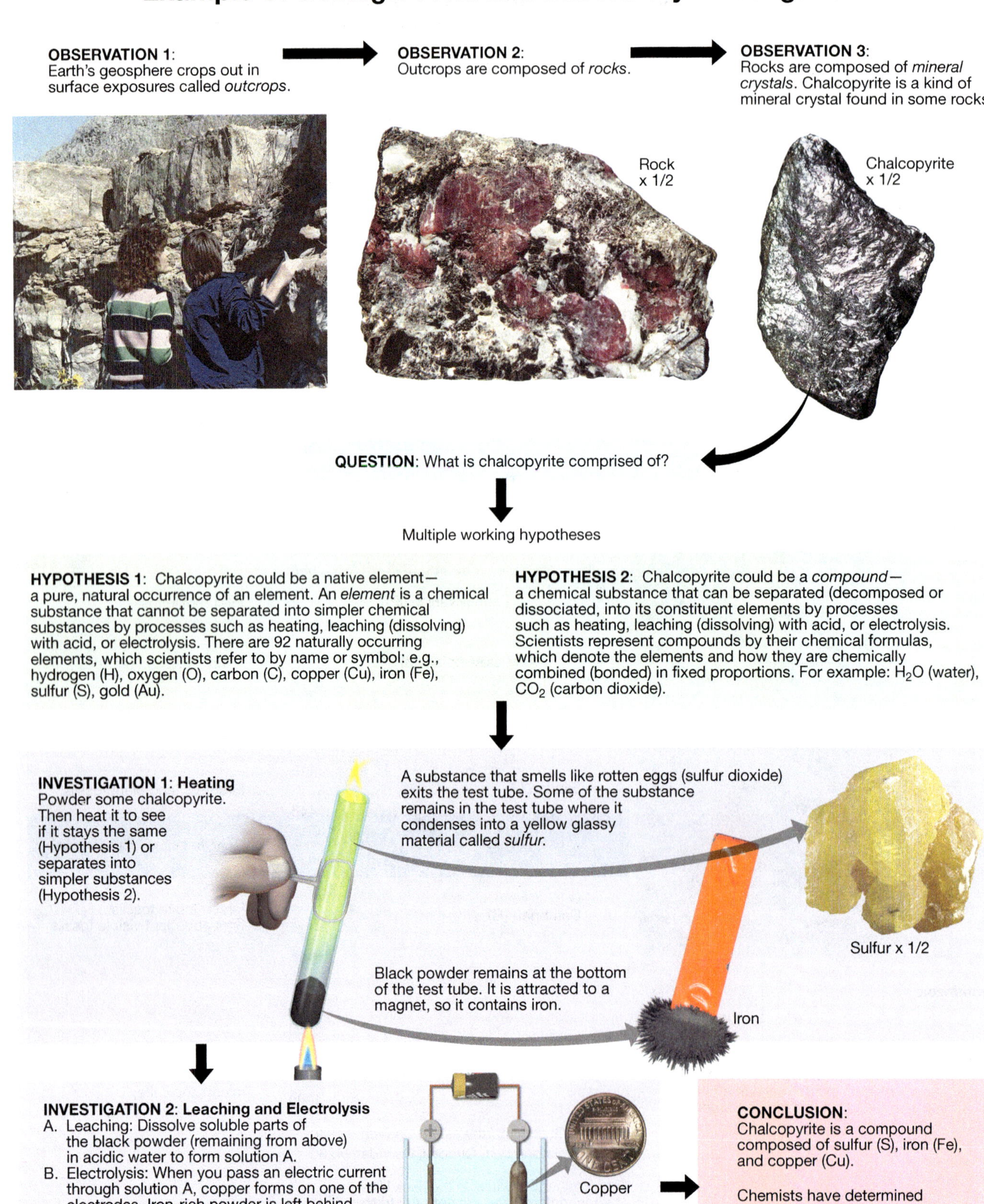

FIGURE 1.4 Example of geologic field and laboratory investigation.

San Andreas Transform-Boundary Plate Motions

Name: ______________________ **Course/Section:** ______________________ **Date:** ____________

Study the geologic map of southern California below, showing the position of the famous San Andreas Fault, a transform plate boundary between the North American Plate (east side) and the Pacific Plate (west side). It is well known to all who live in southern California that plate displacements along the fault cause frequent earthquakes, which place humans and their properties at risk.

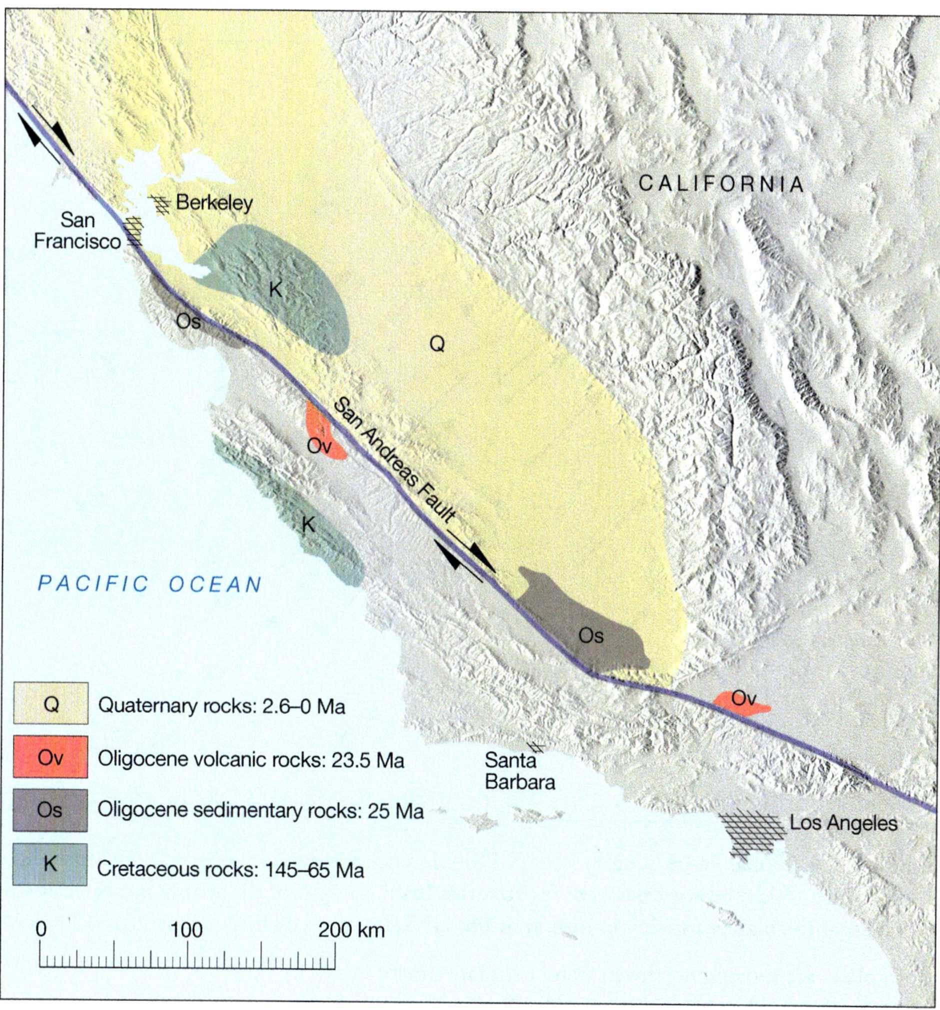

A. The two bodies of Oligocene volcanic rocks (about 23.5 million years old) located along either side of the San Andreas Fault (map above) were once one body of rock, but they have been separated by displacements along the fault. Note that half arrows have been placed along the sides of the fault to show **relative plate motion** along the transform plate boundary here.

1. You can calculate the average annual rate of relative plate motion (displacement) along the San Andreas Fault by measuring how much the Oligocene volcanic rocks have been offset by the fault and by assuming that these rocks began separating soon after they formed. What is the average rate of fault displacement in centimeters per year (cm/yr)? Show your work.

2. An average displacement of about 5 m (16 ft) along the San Andreas Fault was associated with the devastating 1906 San Francisco earthquake that killed people and destroyed properties. Assuming that all displacement along the fault was produced by Earth motions of this magnitude, how often must such earthquakes have occurred in order to account for the total displacement? Show your work.

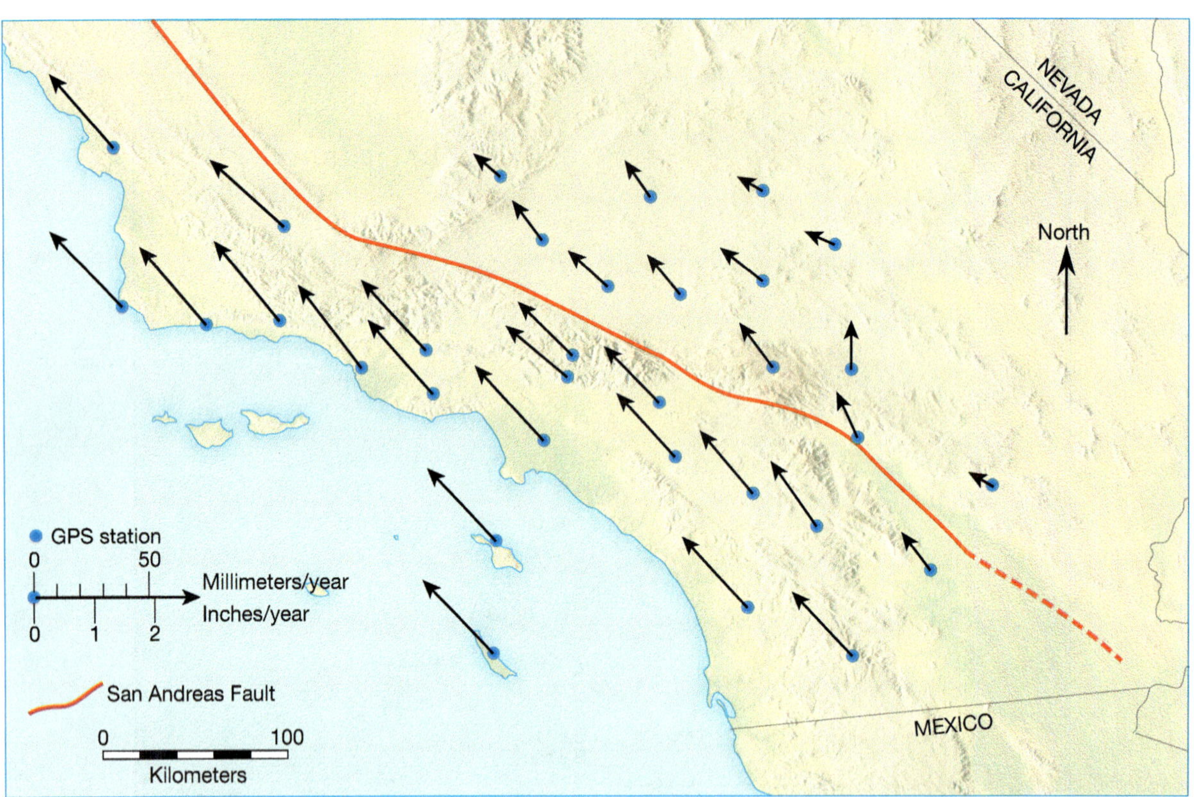

B. The above map shows some Global Positioning System (GPS) reference stations and observations from the JPL-NASA GPS Time Series website at **http://sideshow.jpl.nasa.gov/post/series.html**. Length of the arrows indicates **absolute plate motion**, the direction and rate that the plate is moving in mm/yr at the GPS station (which is attached to bedrock of the plate).

1. Notice that both plates are moving northwest here. Estimate in cm/year how much faster the Pacific Plate is moving than the North American Plate. ____________________ cm/yr
2. Add half arrows along the San Andreas Fault to show the relative movement between the two plates.

C. **REFLECT & DISCUSS** What is the difference between absolute plate motion and relative plate motion?

FLOWCHART FOR CLASSIFICATION OF ROCKS AS IGNEOUS, SEDIMENTARY, OR METAMORPHIC

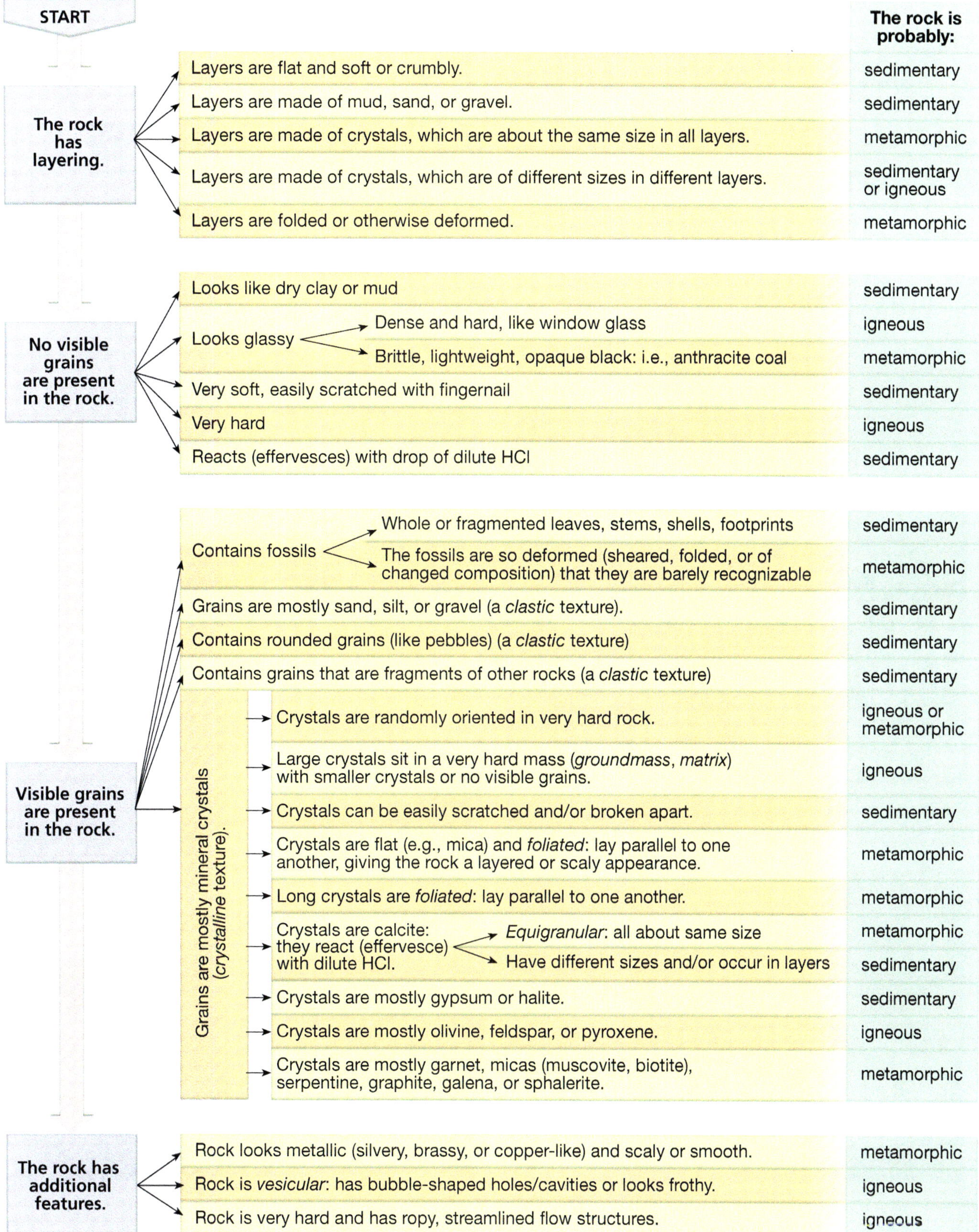

FIGURE 4.4 Flowchart for classification of rocks as igneous, sedimentary, or metamorphic.

FIGURE 4.5 Photograph of a rock sample for analysis, classification, and evaluation. (×1.0)

IGNEOUS ROCK ANALYSIS AND CLASSIFICATION

STEP 1 & 2: MCI and Mineral Composition

Mafic Color Index (MCI): the percent of mafic (green, dark gray, black) minerals in the rock.
See the top of Figure 5.2 and GeoTools Sheets 1 and 2 for tools to visually estimate MCI.

FELSIC MINERALS

Quartz
hard, transparent, gray, crystals with no cleavage

Plagioclase Feldspar
hard, opaque, usually pale gray to white crystals with cleavage, often striated

Potassium Feldspar
hard, opaque, usually pastel orange, pink, or white crystals with exsolution lamellae

Muscovite Mica
flat, pale brown, yellow, or colorless, crystals that scratch easily and split into sheets

MAFIC MINERALS

Biotite Mica
flat, glossy black crystals that scratch easily and split into sheets

Amphibole
hard, dark gray to black, brittle crystals with two cleavages that intersect at 56 and 124 degrees

Pyroxene (augite)
hard, dark green to green-gray crystals with two cleavages that intersect at nearly right angles

Olivine (gemstone peridot)
hard, transparent to opaque, pale yellow-green to dark green crystals with no cleavage

STEP 3: Texture

INTRUSIVE ORIGIN

Pegmatitic
mostly crystals larger than 1 mm: very slow cooling of magma

Phaneritic
crystals about 1–10 mm, can be identified with a hand lens: slow cooling of magma

Porphyritic
large and small crystals: slow, then rapid cooling and/or change in magma viscosity or composition

EXTRUSIVE (VOLCANIC) ORIGIN

Aphanitic
crystals too small to identify with the naked eye or a hand lens; rapid cooling of lava

Glassy
rapid cooling and/or very poor nucleation

Vesicular
like meringue: rapid cooling of gas-charged lava

Vesicular
some bubbles: gas bubbles in lava

Pyroclastic or Fragmental:
particles emitted from volcanoes

STEP 4: Igneous Rock Classification Flowchart

Texture is pegmatitic or phaneritic
- Feldspar > mafic minerals
  - K-spar > Plagioclase
    - quartz present... GRANITE[1,2]
    - no quartz.......... SYENITE[1,2]
  - K-spar < Plagioclase.......... DIORITE[1,2]
- Feldspar < mafic minerals
  - MCI = 45–85.......... GABBRO[1,2]
  - MCI = 85–100 (< 15% felsic minerals).......... PERIDOTITE

Texture is aphanitic and/or vesicular
- felsic (MCI = 0–15) and/or pink, white, or pale brown.......... RHYOLITE[2,3]
- intermediate (MCI = 15–45) and/or green to gray.......... ANDESITE[2,3]
- mafic (MCI ≥ 45) and/or dark gray to black.......... BASALT[2,3]
- mafic with abundant vesicles (resembles a sponge).......... SCORIA
- intermediate or felsic with abundant tiny vesicles—like meringue, floats in water.. PUMICE

Glassy texture.......... OBSIDIAN

(GRANITE through OBSIDIAN: Also refer to Figure 5.2)

Pyroclastic (fragmental) texture
- fragments ≤ 2mm.......... TUFF
- fragments > 2mm.......... VOLCANIC BRECCIA

[1]Add *pegmatite* to end of name if crystals are > 1 cm (e.g., granite-pegmatite).
[2]Add *porphyritic* to front of name when present (e.g., porphyritic granite, porphyritic rhyolite).
[3]Add *vesicular* to front of name when present (e.g., vesicular basalt).

FIGURE 5.4 Igneous rock analysis and classification. Step 1—Estimate the rock's mafic color index (MCI). Step 2—Identify the main rock-forming minerals if the mineral crystals are large enough to do so, and estimate the relative abundance of each mineral (using a Visual Estimation of Percent chart from GeoTools Sheet 1 or 2). Step 3—Identify the texture(s) of the rock. Step 4—Use the Igneous Rock Classification Flowchart to name the rock. Start on the left side of the flowchart, and work toward the right side to the rock name.

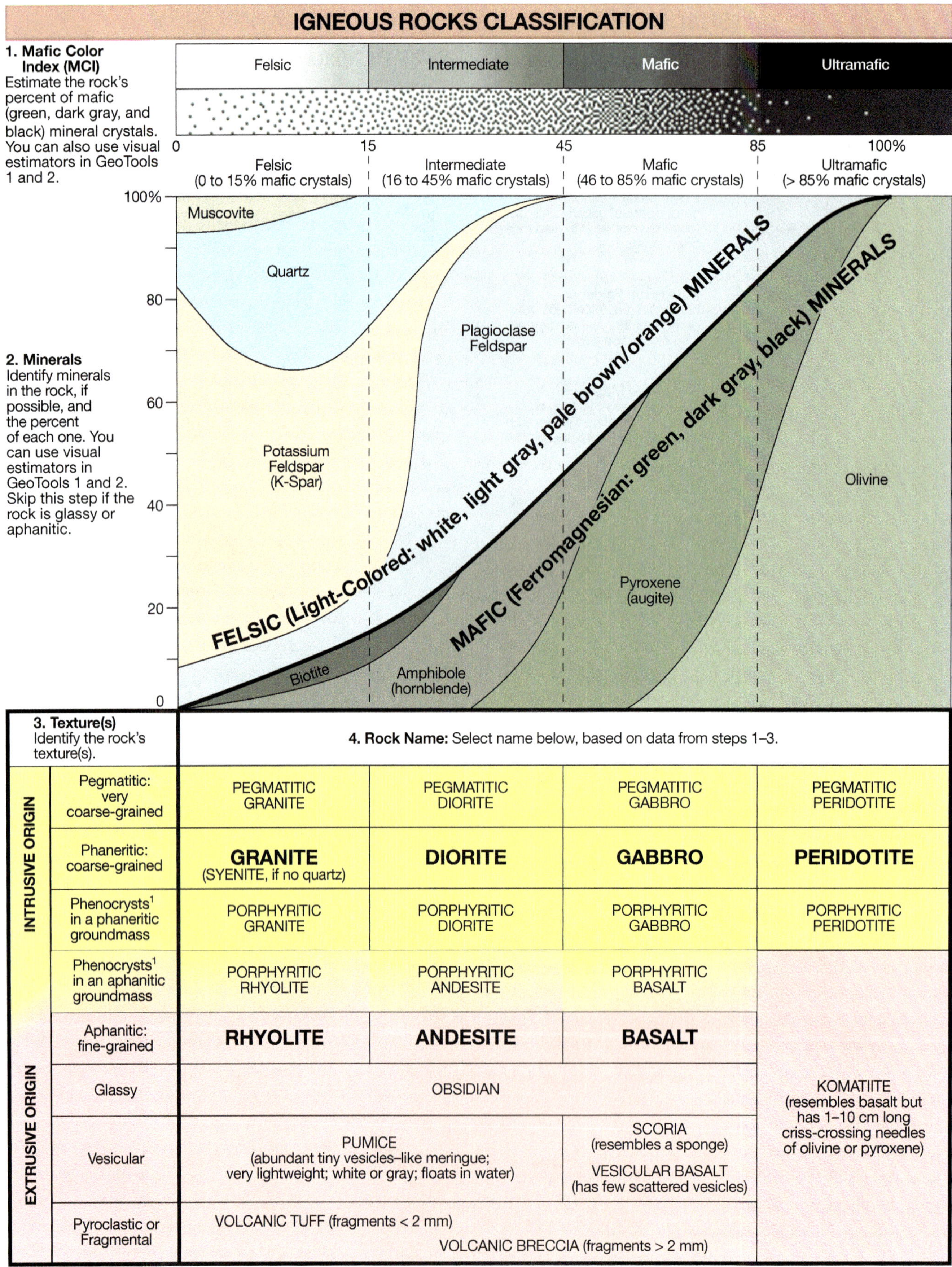

Origin	Texture	Felsic	Intermediate	Mafic	Ultramafic
INTRUSIVE ORIGIN	Pegmatitic: very coarse-grained	PEGMATITIC GRANITE	PEGMATITIC DIORITE	PEGMATITIC GABBRO	PEGMATITIC PERIDOTITE
INTRUSIVE ORIGIN	Phaneritic: coarse-grained	**GRANITE** (SYENITE, if no quartz)	**DIORITE**	**GABBRO**	**PERIDOTITE**
INTRUSIVE ORIGIN	Phenocrysts[1] in a phaneritic groundmass	PORPHYRITIC GRANITE	PORPHYRITIC DIORITE	PORPHYRITIC GABBRO	PORPHYRITIC PERIDOTITE
	Phenocrysts[1] in an aphanitic groundmass	PORPHYRITIC RHYOLITE	PORPHYRITIC ANDESITE	PORPHYRITIC BASALT	KOMATIITE (resembles basalt but has 1–10 cm long criss-crossing needles of olivine or pyroxene)
EXTRUSIVE ORIGIN	Aphanitic: fine-grained	**RHYOLITE**	**ANDESITE**	**BASALT**	
EXTRUSIVE ORIGIN	Glassy	OBSIDIAN			
EXTRUSIVE ORIGIN	Vesicular	PUMICE (abundant tiny vesicles–like meringue; very lightweight; white or gray; floats in water)		SCORIA (resembles a sponge) VESICULAR BASALT (has few scattered vesicles)	
EXTRUSIVE ORIGIN	Pyroclastic or Fragmental	VOLCANIC TUFF (fragments < 2 mm) VOLCANIC BRECCIA (fragments > 2 mm)			

[1]Phenocrysts are crystals conspicuously larger than the finer grained groundmass (main mass, matrix) of the rock.

FIGURE 5.5 Igneous Rock Classification Chart. Obtain data about the rock in Steps 1–3, then use that data to select the name of the rock (Step 4). Also refer to **FIGURE 5.4** and the examples of classified igneous rocks in **FIGURES 5.8–5.14**.

5.6 Rock Analysis, Classification, and Origin

Name: ______________________ **Course/Section:** ______________________ **Date:** __________

A. Analyze and classify each igneous rock below, then infer the origin of each rock based on its texture. Refer to **FIGURE 5.4** and **5.5** as needed. The rocks are shown ×1 (actual size).

Mafic color index (MCI, percentage of mafic mineral crystals): Would you describe the rock as mafic, intermediate, or felsic?	Mafic color index (MCI, percentage of mafic mineral crystals): Would you describe the rock as mafic, intermediate, or felsic?	Mafic color index (MCI, percentage of mafic mineral crystals): Would you describe the rock as mafic, intermediate, or felsic?	Mafic color index (MCI, percentage of mafic mineral crystals): Would you describe the rock as mafic, intermediate, or felsic?
Texture(s) present:	Texture(s) present:	Texture(s) present:	Name and percent abundance of mineral crystals: Texture present:
The name of this rock is:	The name of this rock is:	The name of this rock is:	The name of this rock is:
Based on its texture, how did this rock form?	Based on its texture, how did this rock form?	Based on its texture, how did this rock form?	Based on its texture, how did this rock form?

Bowen's Reaction Series

When magma intrudes Earth's crust, it cools into a mass of mineral crystals and/or glass. Yet when geologists observe and analyze the igneous rocks in a single dike, sill, or batholith, they often find that it contains more than just one kind of igneous rock. Apparently, more than one kind of igneous rock can form from a single homogeneous body of magma as it cools. American geologist, Norman L. Bowen made such observations in the early 1900s. He then devised and carried out laboratory experiments to study how magmas might evolve in ways that could explain how more than one kind of igneous rock could form from a single body of magma. His work is commonly summarized in a diagram (FIGURE 5.6) called **Bowen's Reaction Series**, which shows how different kinds of igneous rocks can evolve from a single body of magma as it cools.

Bowen's Experiment

Other geologic investigations had already suggested that the top of Earth's mantle is made of peridotite. So Bowen placed pieces of peridotite into *bombs*, strong pressurized ovens used to melt the rocks at high temperatures (1200–1400°C). Once the rocks had melted to form peridotite magma, Bowen would allow the magma to cool to a given temperature and remain at that temperature for a while in hopes of having it begin to crystallize. The rock was then quickly removed from the bomb and quenched (cooled by dunking it in water) to make any remaining molten rock form glass. Bowen then identified the mineral crystals that had formed at each temperature. His experiments showed that as magma cools in an otherwise unchanging environment, two series of silicate minerals crystallize in a predictable order.

Discontinuous Crystallization of Mafic Minerals (Left Branch). The left branch of Bowen's Reaction Series (FIGURE 5.6) shows the predictable series of mafic minerals that crystallize from a peridotite magma that is allowed to cool slowly. This series is discontinuous because one mafic mineral replaces another as the magma cools. For example, olivine is first to crystallize at very high temperature. But if the magma cools to about 1100° C, then the olivine starts to react with it and dissolve as pyroxene (next mineral in the series) starts to crystallize. More cooling of the magma causes pyroxene to react with the magma as amphibole (next mineral in the series) starts to crystallize, and so on. If the magma cools too quickly, then rock can form while one reaction is in progress and before any remaining reactions even have time to start.

Continuous Crystallization of Plagioclase (Right Branch). The right branch of Bowen's Reaction Series (FIGURE 5.6) shows that plagioclase feldspar crystallizes continuously from high to low temperatures (~1100–800° C), but this is accompanied by a series of continuous change in the composition of the plagioclase. The high temperature plagioclase is calcium rich and sodium poor, and the low temperature plagioclase is sodium rich and calcium poor. If the magma cools too quickly for the plagioclase to react with the magma, then a single plagioclase crystal can have a more calcium rich center and a more sodium rich rim.

Crystallization of Quartz (Bottom of the Series). Finally, notice what happens at the bottom of Bowen's Reaction Series (FIGURE 5.6). At the lowest temperatures, where the last crystallization of magma occurs, the remaining elements form abundant potassium feldspar (K-spar), muscovite, and quartz.

Partial Melting and Bowen's Reaction Series. When a plastic tray of ice cubes is heated in an oven, the ice cubes melt long before the plastic tray melts (i.e., the ice cubes melt at a much lower temperature). As rocks are heated, their different mineral crystals also melt at different temperatures. Therefore, at a given temperature, it is possible to have rocks that are partly molten and partly solid. This phenomenon is known as *partial melting*. When minerals of Bowen's Reaction Series are heated, they melt at different temperatures. The plagioclase feldspars melt continuously from about 1100–1500° C, but the ferromagnesian minerals, quartz, and K-feldspar melt discontinuously. K-feldspar melts at about 1250° C, pyroxene at 1400° C, quartz at 1650° C, and olivine at 1800° C. Because feldspars tend to melt at lower temperatures than the ferromagnesian minerals, partial melting of an igneous rock tends to produce magma of more felsic composition than the original rock from which it melted. So when a rock like basalt partially melts, it tends to form a magma that is more felsic and would cool to form andesite.

Magmatic Differentiation. Bowen's Reaction Series is an example of one way that more than one rock type can form from a single body of magma. It was generated under controlled laboratory conditions. There is no known natural location where an ultramafic magma evolved to a felsic one according to Bowen's Reaction Series. However, there are many examples where parts of Bowen's Reaction Series have occurred in nature.

Bowen's continuous series of crystallization leads to the depletion of calcium and sodium from the magma, so the composition of the magma changes. However, along the discontinuous series, early-formed mafic mineral crystals in a cooling body of magma have been shown to react with the magma at lower temperatures to form new mafic minerals. If this recycling of elements occurred perfectly, then the concentrations of iron and magnesium in the magma would never change. In nature, some of the early-formed crystals either settle out of the magma or are encrusted with different minerals before they can react, so they can no longer react with the original magma. This is called *fractional crystallization*. On the other hand, a magma may melt some of the wall rocks surrounding it and assimilate its elements. This is called *assimilation*. *Magma mixing* may also occur. Bowen's continuous series of crystallization, fractional crystallization, assimilation, and magma mixing are all factors that can contribute to **magmatic differentiation** (any process that causes magma composition to change). Magmatic differentiation produces more than one rock type from a single body of magma.

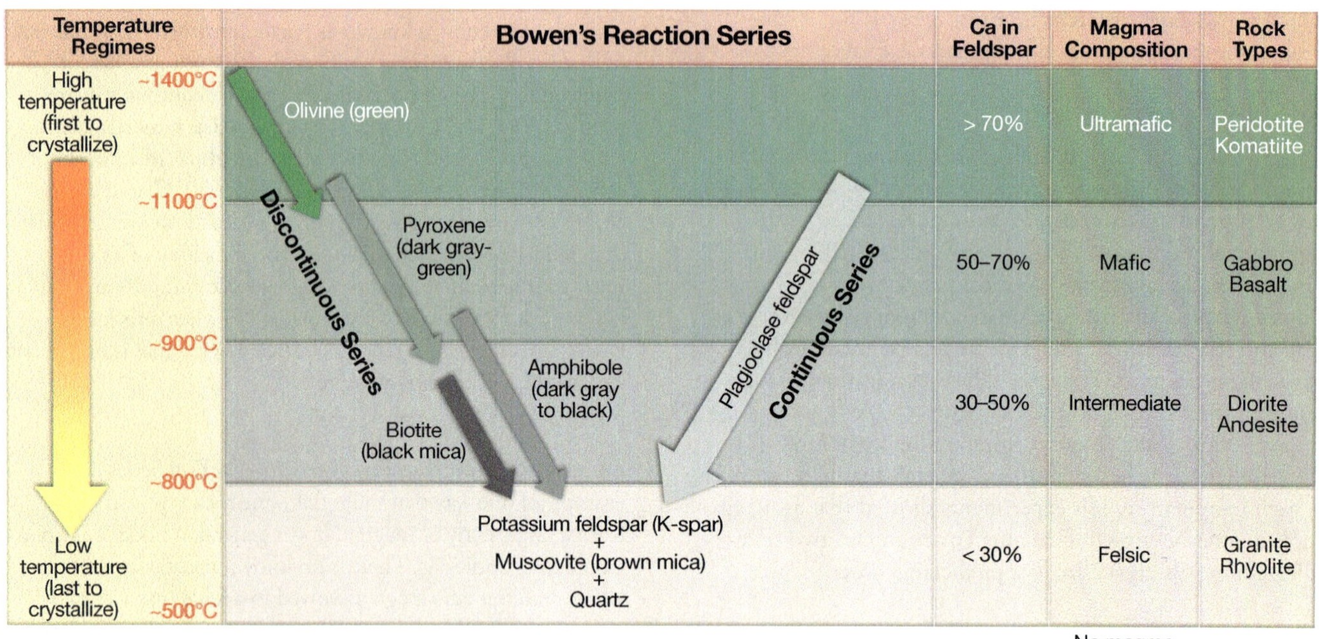

FIGURE 5.6 Bowen's Reaction Series—A laboratory-based conceptual model of one way that different kinds of igneous rocks can differentiate from a single, homogeneous body of magma as it cools. See text for discussion.

5.7 Thin Section Analysis and Bowen's Reaction Series

Name: ______________________ **Course/Section:** ______________ **Date:** ________

A thin section of a rock can be made by grinding one side of it flat, gluing the flat side to a glass slide, and then grinding the rock so thin that light passes through it. The thin section is then viewed with a polarizing microscope. The view in plane polarized light is the same as looking at the thin section through a pair of sunglasses. If you place a second pair of sunglasses behind the thin section and hold it perpendicular to the first pair, then you are viewing the thin section through cross polarized light. These images of thin sections were made by geologist LeeAnn Srogi, West Chester University of Pennsylvania.

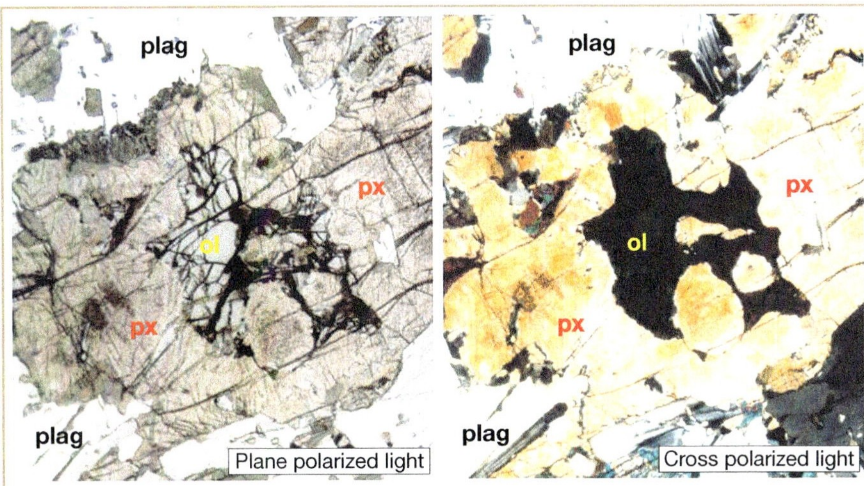

Thin section 1: This thin section has a crystal of olivine (ol) that is medium gray in plane polarized light and black in cross polarized light. The olivine crystal once had a rectangular outline, but is now surrounded and partly replaced by the mineral pyroxene (px), which has a different composition and crystal structure. Pyroxene is pale brown in plane polarized light, but yellow- to orange-brown in cross polarized light. The other white to light gray crystals in this rock are plagioclase (plag).

← 1 mm →

Thin section 2: This thin section is shown in cross polarized light. There are crystals of amphibole (brown to green color) and plagioclase (white to gray color). The large plagioclase crystal in the center of the image is "zoned." It has a calcium rich (sodium poor) center surrounded by zones that also have progressively more and more sodium. The zone at its outer edge is equally rich in both calcium and sodium.

1 mm

A Based on Bowen's Reaction Series (**FIGURE 5.6**), explain as exactly as you can what may have caused the relationship between olivine and pyroxene observed in thin section 1.

B Based on Bowen's Reaction Series (**FIGURE 5.6**), explain as exactly as you can what may have caused the large plagioclase crystal in thin section 2 to be zoned as it is.

C. REFLECT & DISCUSS Based on your work above, circle and label the parts of this Bowen's Reaction Series diagram to indicate the exact path of crystallization and reaction represented in thin section 1, and then 2. Refer to **FIGURE 5.6** as needed.

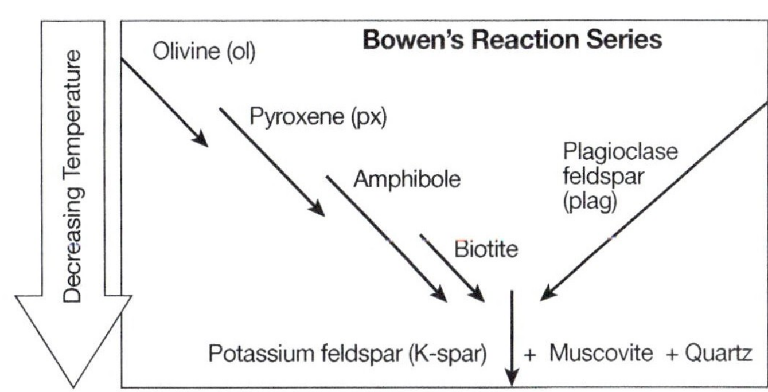

COMPOSITIONAL CLASSIFICATION OF SEDIMENT AND SEDIMENTARY ROCKS

A. DETRITAL (SILICICLASTIC) SEDIMENT AND SEDIMENTARY ROCK IS MOSTLY ONE OR MORE OF THESE:

Rock fragments: may be angular or rounded; can include detrital chert grains (see "chert" below)

Quartz grains: angular grains freshly broken from their source and pebbles rounded during transportation

Feldspar grains: large angular grains freshly broken from their source and small subangular grains

Clay: commonly forms from chemical decay of feldspars and micas

B. BIOCHEMICAL SEDIMENT AND SEDIMENTARY ROCK IS MOSTLY EITHER OR BOTH OF THESE:

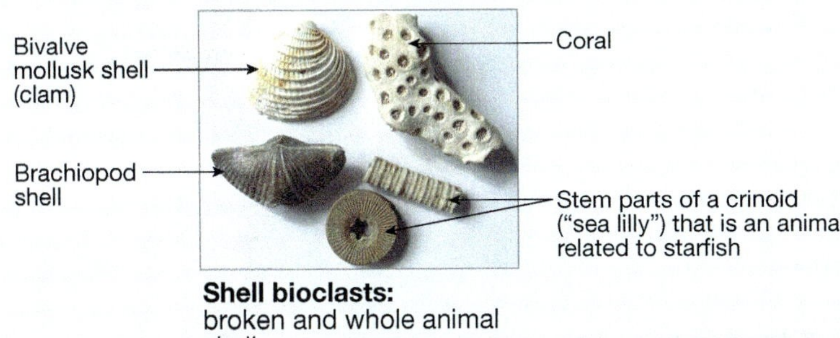

Shell bioclasts: broken and whole animal shells

Plant fragments: are brown in peat and black in coal

C. CHEMICAL SEDIMENT AND SEDIMENTARY ROCK IS MOSTLY MADE OF ONE OR MORE OF THESE:

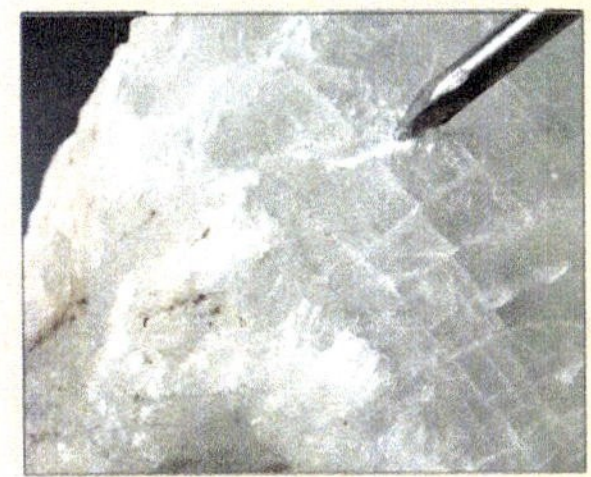

Gypsum: white or gray, easily scratched with your fingernail

Calcite spar (crystals): reacts with dilute HCl, breaks into rhombohedral shapes

Dolomite: usually cryptocrystalline; reacts with dilute HCl only if it is powdered

Halite: gray to red cubic crystals (often intergrown as rock salt); salty taste

Ooids: tiny (< 2 mm) spheres of calcite or aragonite that resemble miniature pearls; reacts to dilute HCl

Limonite: opaque brown to yellow rusty-looking crusts, layers; cements sediment, making it look yellow to brown

Hematite: opaque brick red to silver gray layers; cements sediment, making it look red

Chert: a gray, red, brown or black cryptocrtstalline variety of quartz (may contain fossils, including silica microfossils

FIGURE 6.2 Composition of sedimentary rocks. Scale for all images is × 1 unless noted otherwise.

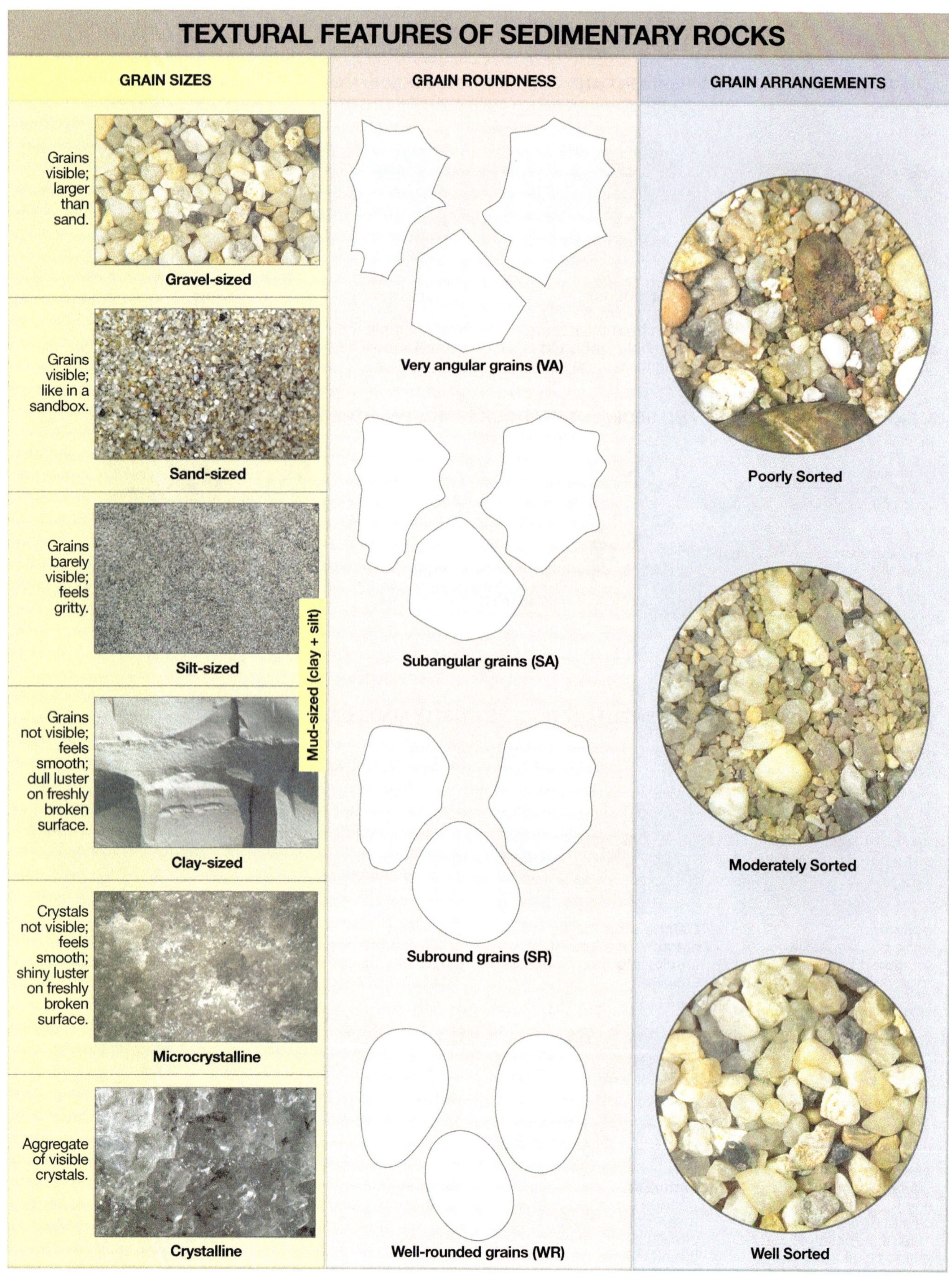

FIGURE 6.3 Textures of sedimentary rocks. Scale for all images is × 1.

SEDIMENTARY ROCK ANALYSIS AND CLASSIFICATION

STEP 1: Composition. What materials comprise most of the rock?		STEP 2: What are the rock's texture and other distinctive properties?		STEP 3: Name the rock based on your analysis in steps 1 and 2.		
Detrital (Siliciclastic) sediment grains: fragmented rocks and/or silicate mineral crystals	Rock fragments and/or quartz grains and/or feldspar grains and/or clay minerals (e.g., kaolinite) Detrital sediment is derived from the mechanical and chemical weathering of continental (land) rocks, which consist mostly of silicate minerals. Detrital sediment is also called terrigenous (land derived) sediment.	Mostly angular and/or subangular gravel (grains larger than 2 mm)		BRECCIA*		Detrital (Siliciclastic) sedimentary rocks
		Mostly subround and/or well rounded gravel (grains larger than 2 mm)		CONGLOMERATE*		
		Mostly sand (1/16–2 mm grains). May contain fossils	Mostly quartz sand	QUARTZ SANDSTONE	SANDSTONE	
			Mostly feldspar sand	ARKOSE		
			Mostly rock fragment sand	LITHIC SANDSTONE		
			Sand is mixed with much mud	WACKE (GRAYWACKE)		
		Mud (< 1/16 mm): Mostly silt. May contain fossils	Breaks into blocks or layers	SILTSTONE	MUDSTONE	
		No visible grains; Mud (< 1/16 mm): Mostly clay. May contain fossils	Fissile (splits easily into layers)	SHALE		
			Crumbles into blocks	CLAYSTONE		
Biochemical (Bioclastic) sediment grains: fragments/shells of organisms	Plant fragments and/or charcoal	Brown porous rock with visible plant fragments that are easily broken apart from one another		PEAT		Biochemical (Bioclastic) sedimentary rocks
		Dull, dark brown, brittle rock; fossil plant fragments may be visible		LIGNITE		
		Black, layered, brittle rock; may be sooty or bright		BITUMINOUS COAL		
	Shells and shell/coral fragments, and/or calcareous microfossils	Mostly gravel-sized shells and shell or coral fragments; (Figure 6.6)		COQUINA	LIMESTONE	
		Mostly sand-sized shell fragments; often contains a few larger whole fossil shells		CALCARENITE (FOSSILIFEROUS LIMESTONE)		
		Silty, earthy rock comprised of the microscopic shells of calcareous phytoplankton (microfossils); may contain a few visible fossils		CHALK		
		No visible grains: No visible grains in most of the rock. May break with conchoidal fracture. May contain a few visible fossils in the micrite		MICRITE		
Mineral crystals (inorganic) or chemical residues (e.g., rust)	Calcite crystals and/or calcite spheres and/or microcrystalline calcite/aragonite	Mostly spherical grains that resemble miniature pearls (< 2 mm), called ooliths or ooids		OOLITIC LIMESTONE	LIMESTONE	Chemical sedimentary rocks
		Masses of visible crystals and/or microcrystalline; may have cavities, pores, or color banding (Figure 6.8); usually light colored		TRAVERTINE		
	Microcrystalline dolomite	Effervesces in dilute HCl only if powdered. Usually light colored. (Commonly forms from alteration of limestone)		DOLOSTONE		
	Halite mineral crystals	Visible cubic crystals, translucent, salty taste (Figure 6.7)		ROCK SALT		
	Gypsum mineral crystals	Gray, white, or colorless. Visible crystals or microcrystalline. Can be scratched with your fingernail		ROCK GYPSUM		
	Iron-bearing minerals crystals or residues	Dark-colored, dense, amorphous masses (e.g. limonite), microcrystalline nodules or inter-layered with quartz or red chert (banded iron formation)		IRONSTONE		
	Microcrystalline varieties of quartz (flint, chalcedony, chert, jasper)	Microcrystalline, may break with a conchoidal fracture. Hard (scratches glass). Usually gray, brown, black, or mottled mixture of those colors. Chert can be regarded as biochemical if its silica came from dissolution of siliceous plankton (diatoms, radiolaria).		CHERT (a siliceous rock)		

rocks that effervesce in dilute HCl (Coquina through Travertine)

carbonate/calcareous rocks (Coquina through Dolostone)

evaporite rocks (Travertine through Rock Gypsum)

clastic rocks (Detrital and Biochemical sedimentary rocks)

*Modify name as quartz breccia/conglomerate, arkose breccia/conglomerate, lithic breccia/conglomerate, or wacke breccia/conglomerate as done for sandstones.

FIGURE 6.9 Sedimentary rock analysis and classification. See page 166 for steps to analyze and name a sedimentary rock.

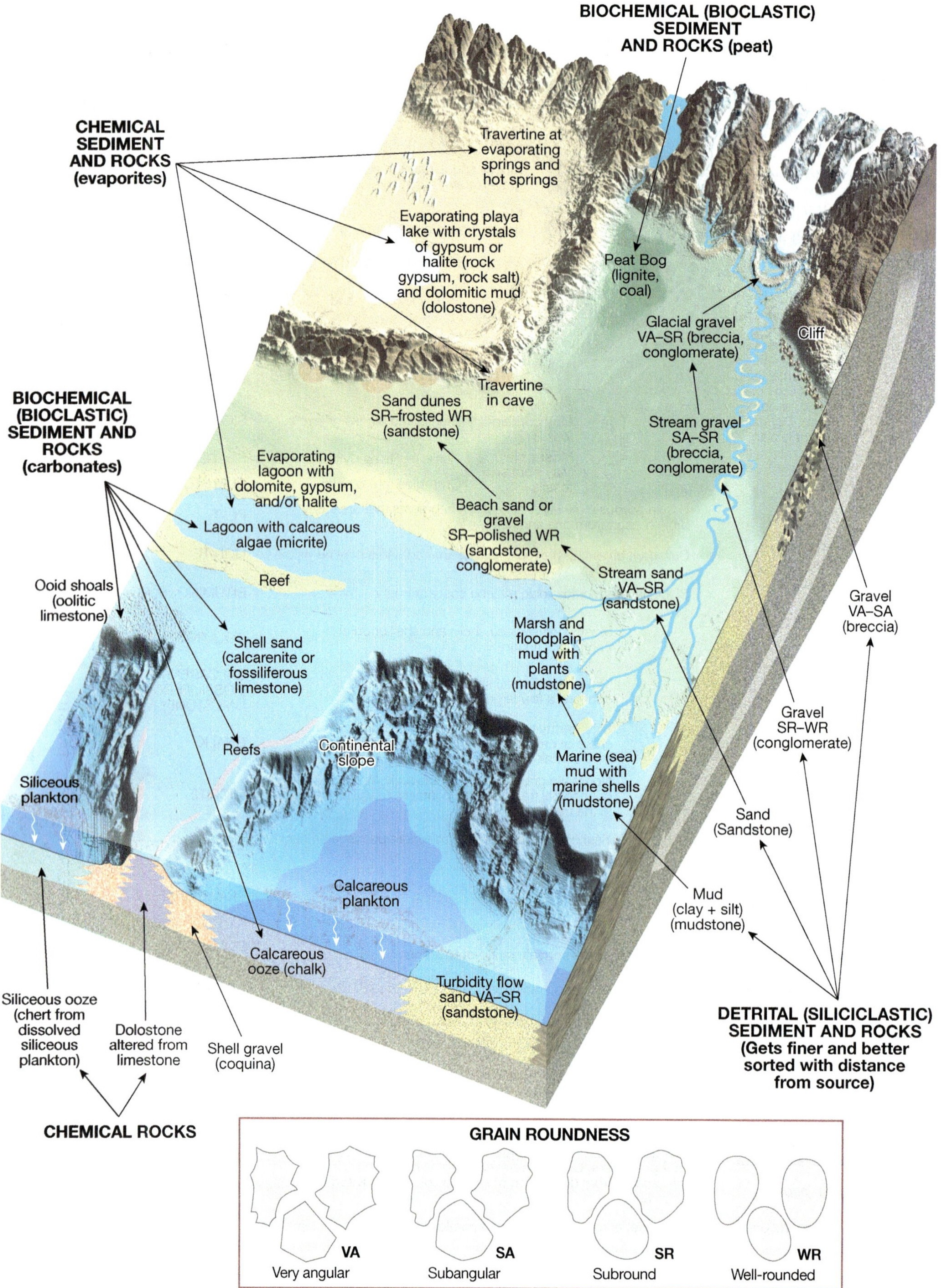

FIGURE 6.10 Sedimentary environments. Some named modern environments where specific kinds of sediments and sedimentary rocks are forming.

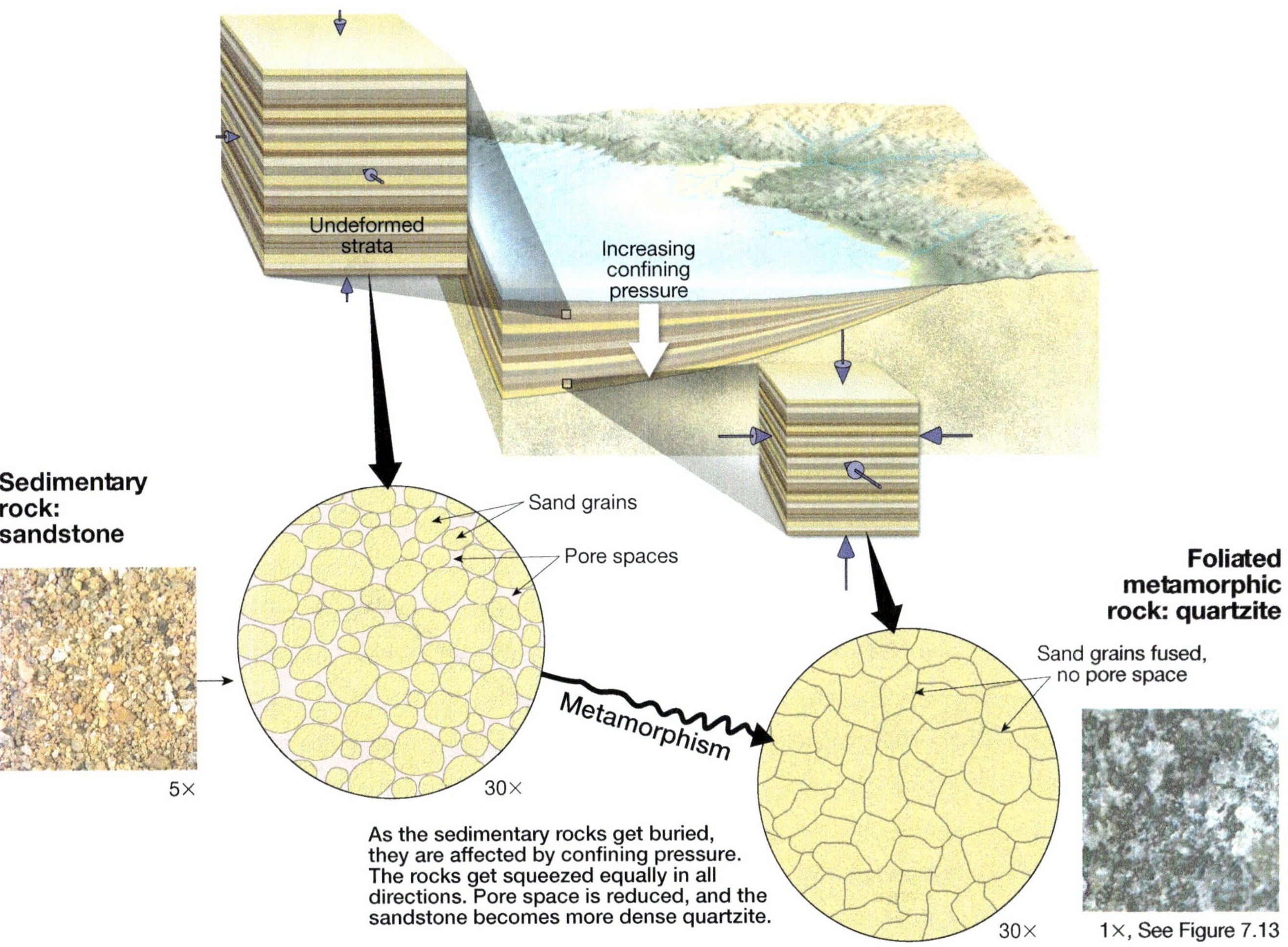

FIGURE 7.2 Effects of confining pressure (equal stress). As rocks get buried, they experience confining pressure that is equal in all directions (equal stress). The rocks become more dense as pore space is squeezed, may recrystallize to crystals of equal size, and remain nonfoliated.

EFFECTS OF DIRECTED PRESSURE (DIFFERENTIAL STRESS)

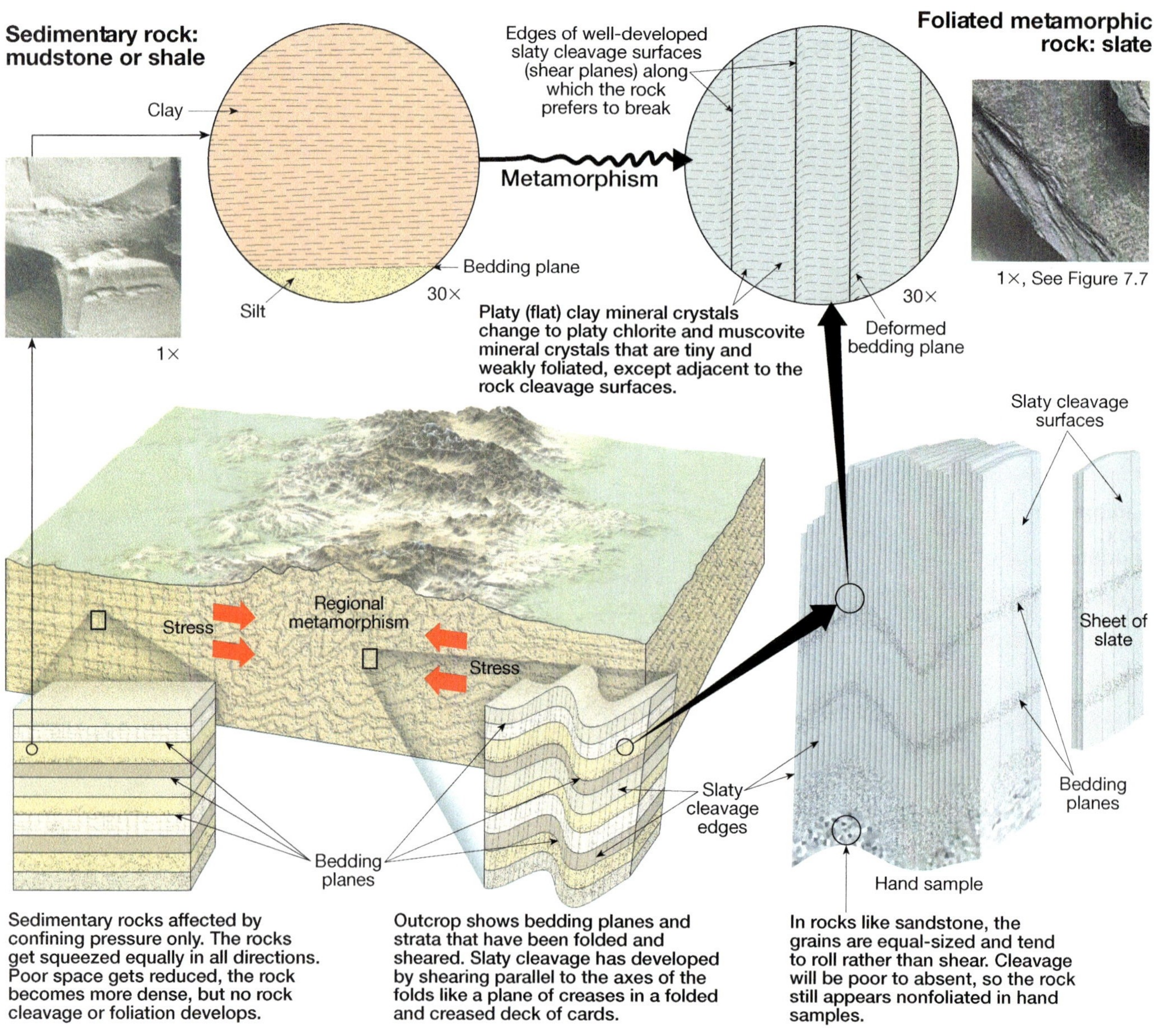

FIGURE 7.3 Effects of directed pressure (unequal stress). In places like convergent plate boundaries, rocks experience directed pressure, also known as *differential stress*, as they collide. They may fracture (brittle rocks) or fold (ductile rocks) and develop rock cleavage and foliation of platy (flat) minerals.

METAMORPHIC ROCK ANALYSIS AND CLASSIFICATION

STEP 1: What are the rock's textural features?			STEP 2: What are the rock's mineralogical composition and/or other distinctive features?	STEP 3: Metamorphic rock name	STEP 4: What was the parent rock?	STEP 5: What is the rock used for?
FOLIATED	Fine-grained or no visible grains	Flat slaty cleavage is well developed	Dull luster; breaks into hard flat sheets along the slaty cleavage	SLATE[1]	Mudstone or shale	Roofing slate, table tops, floor tile, and blackboards
FOLIATED	Fine-grained or no visible grains	Phyllite texture well developed more than slaty cleavage	Breaks along wrinkled or wavy foliation surfaces with shiny metallic luster	PHYLLITE[1]	Mudstone, shale, or slate	Construction stone, decorative stone, sources of gemstones
FOLIATED	Medium- to coarse-grained	Schistosity: foliation formed by alignment of visible crystals; rock breaks along scaly foliation surfaces; crystalline texture	Mostly blue or violet needle-like crystals (blue amphibole)	Blueschist (SCHIST[1])	Mudstone, shale, slate, or phyllite	
			Mostly visible sparkling crystals of chlorite +/– actinolite (green amphibole)	Greenschist (SCHIST[1])		
			Mostly visible sparkling crystals of muscovite	Muscovite schist (SCHIST[1])		
			Mostly visible sparkling crystals of biotite	Biotite schist (SCHIST[1])		
FOLIATED	Medium- to coarse-grained	Gneissic banding: minerals segregated into alternating layers gives the rock a banded texture in side view; crystalline texture	Visible crystals of two or more minerals in alternating light and dark foliated layers	GNEISS[1]	Mudstone, shale, phyllite, schist, granite, or diorite	Construction stone, decorative stone, sources of gemstones
FOLIATED OR NONFOLIATED		Medium- to coarse-grained crystalline texture	Mostly visible glossy black amphibole (hornblende) in blade-like crystals	AMPHIBOLITE	Basalt, gabbro, or ultramafic igneous rocks	Construction stone
FOLIATED OR NONFOLIATED		Crystalline texture	Green pyroxene + red garnet	ECLOGITE	Basalt, gabbro	Titanium ore
NONFOLIATED	Fine-grained or no visible grains	Glassy texture; slaty cleavage may barely be visible	Black glossy rock that breaks along uneven or conchoidal fractures (Figure 7.12)	ANTHRACITE COAL	Peat, lignite, bituminous coal	Highest grade coal for clean burning fossil fuel
NONFOLIATED	Fine-grained or no visible grains	Microcrystalline texture	Usually a dull dark color; very hard	HORNFELS	Any rock type	
NONFOLIATED	Fine-grained or no visible grains	Microcrystalline texture or no visible grains. May have fibrous asbestos form	Serpentine; dull or glossy; color usually shades of green	SERPENTINITE	Basalt, gabbro, or ultramafic igneous rocks	Decorative stone
NONFOLIATED	Fine-grained or no visible grains	Microcrystalline or no visible grains	Talc; can be scratched with your fingernail; shades of green, gray, brown, white	SOAPSTONE	Basalt, gabbro, or ultramafic igneous rocks	Art carvings, electrical insulators, talcum powder
NONFOLIATED	Fine- to coarse-grained	Sandy texture	Quartz sand grains fused together; grains will not rub off like sandstone; usually light colored	QUARTZITE[1]	Sandstone	Construction stone, decorative stone
NONFOLIATED	Fine- to coarse-grained	Microcrystalline (resembling a sugar cube) or medium to coarse crystalline texture	Calcite (or dolomite) crystals of nearly equal size and tightly fused together; calcite effervesces in dilute HCl; dolomite effervesces only if powdered	MARBLE[1]	Limestone	Art carvings, construction stone, decorative stone, source of lime for agriculture
NONFOLIATED	Fine- to coarse-grained	Conglomeratic texture, but breaks across grains	Pebbles may be stretched or cut by rock cleavage	META-CONGLOMERATE	Conglomerate	Construction stone, decorative stone

Step 3 arrow (SLATE through GNEISS): INCREASING METAMORPHIC GRADE

[1]Modify rock name by adding names of minerals in order of increasing abundance. For example, garnet muscovite schist is a muscovite schist with a small amount of garnet.

FIGURE 7.16 Five-step chart for metamorphic rock analysis and classification. See text for description of steps (page 198).

9780134984490

TP
APEX